Copyright © 2021 by Deanne Dietz

All rights reserved. No part of this publication may be reproduced, stored or transmitted in any form or by any means, electronic, mechanical, photocopying, recording, scanning, or otherwise without written permission from the publisher. It is illegal to copy tl1is book, post it to a website, or distribute it by any other means without permission.

Deanne Dietz asserts the moral right to be identified as the author of this work.

Designations used by companies to distinguish their products are often claimed as trademarks. All brandnames and product names used in this book and on its cover are trade names, service marks, trademarks and registered trademarks of their respective owners. The publishers and the book are not associated with any product or vendor mentioned in this book. None of the companies referenced within the book have endorsed the book.

**first edition**

This book was professionally typeset on Reedsy.
**Find out more** at reedsy.com

The Teeny Tiny Telescope soared high in the sky, as they passed their new friend and said, "Goodbye!"

The Teeny Tiny Telescope stared off into the quiet nearby and spotted a planet off to the side.

The Teeny Tiny Telescope rocketed towards the oncoming horizon.

As they stated, "Hello," to the sleepy planet.

The Teeny Tiny Telescope was taken aback by the planets bright glow,

The Teeny Tiny Telescope listened as Venus replied, "I was named after the God of beauty and love."

The Teeny Tiny Telescope admired Venus's brightness and asked to venture into their atmosphere. "Of course!" They replied.

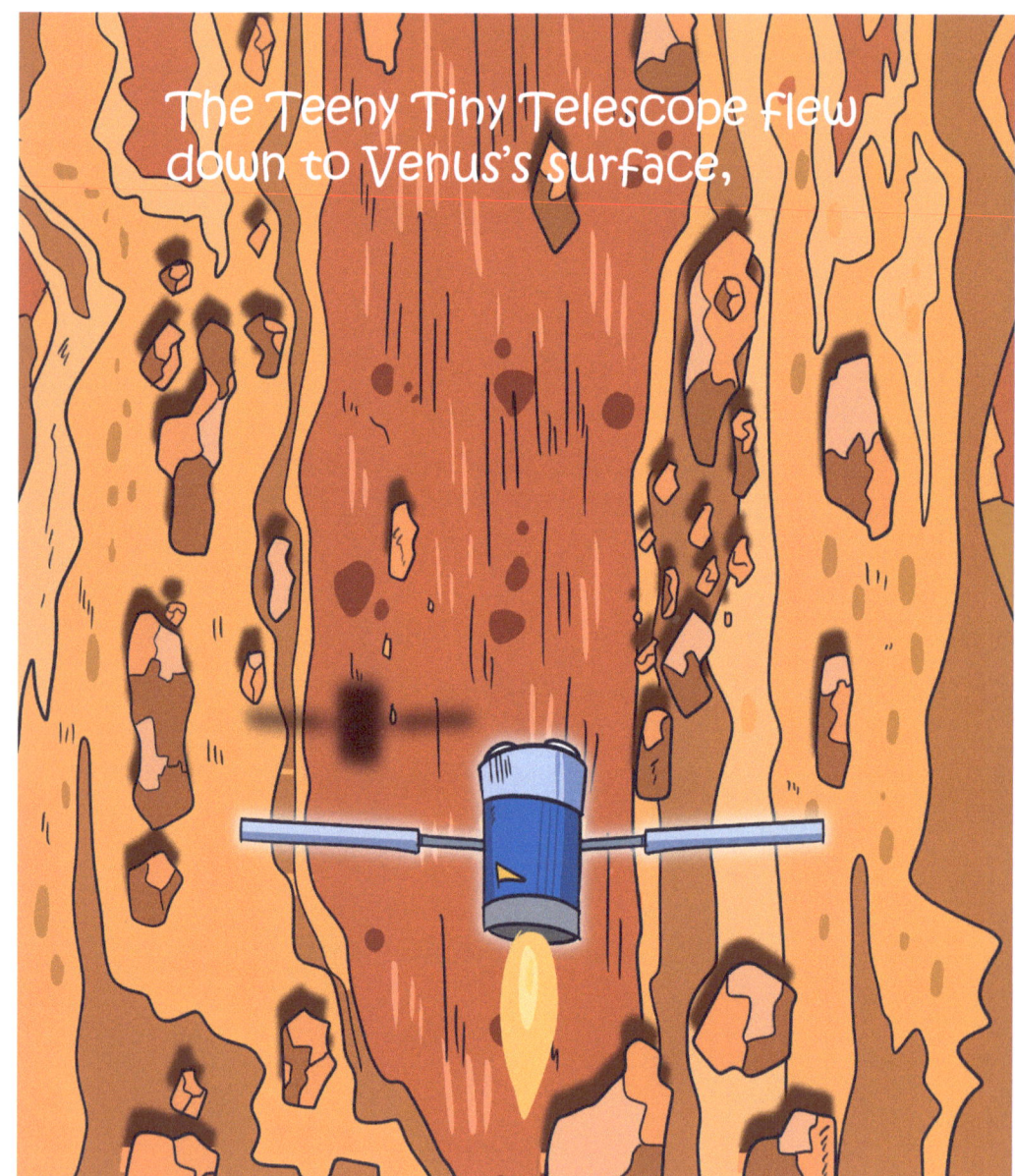

The Teeny Tiny Telescope quickly realized Venus's volcanoes and clouds cause it to be the hottest planet in the sky.

The planet stated, "I have none nearby."

The Teeny Tiny Telescope thanked them as they said, "Goodbye!"

www.ingramcontent.com/pod-product-compliance
Lightning Source LLC
Chambersburg PA
CBHW040348220526

45473CB00009B/2818